SOTHIS PRESS

Sothis Press aims to completely reimagine the public domain book with an emphasis on universal access to captivatingly designed print editions. We believe that technology will enable every single book ever published to be perpetually accessible in beautiful print form. Sothis is beauty in ubiquity.

www.sothis.press

1 Waterhouse Square, London, ECIN 2ST

The Perfection Method of Wood Steaming

Live High Pressure Steam Vs. Low Pressure, Its Application to Wood for Bending

Perfection Wood Steaming Retort Co

SOTHIS PRESS

ISBN 978-1-0365-2496-8

The Perfection Method of Wood Steaming is in the public domain.

This edition is published by Sothis Press © 2023. The company is registered in England and Wales. The Company Number is 14900389.

The PERFECTION METHOD OF WOOD STEAMING

Live High Pressure Steam
***vs* Low Pressure**

Its Application to Wood for Bending

PERFECTION WOOD STEAMING
RETORT CO., Parkersburg, W. Va.

IT IS with pleasure that we place this catalog in the hands of our friends of the woodworking trades. It explains clearly the "Perfection" method of steaming stock for bending purposes.

Up to the time the method of steaming stock under pressure was put in use by us in 1907 and the "Perfection" Retorts were patented, there had never been any improvements made in the steaming of Wood for bending purposes.

Before "Perfection" Retorts were put on the market, steaming in open boxes or vats with exhaust steam or by boiling water were in vogue. These old methods are fast giving way to the new and scientific "Perfection" way.

PERFECTION WOOD STEAMING RETORT CO.
PARKERSBURG, WEST VIRGINIA

Live High Pressure Steam
vs
Low Pressure

TO the practical mind it must appear that because of the conditions imposed and the great volume required, the use of exhaust steam in the treatment of wood for bending is wasteful. It is a fact well known to those familiar with the subject that exhaust steam, because of retained latent heat, is of practically the same heating value as an equal quantity of direct steam. As a solutive in its application to wood for bending, it is of relatively little value owing to the fact that it cannot, after its escape from the engine, be confined as a force. It must be allowed free circulation and final unobstructed exit to the atmosphere, or subjected to rapid condensation if retained, to avoid back pressure on the piston and consequent loss of power.

The disintegrating influence of the escaping steam and dripping water on the buildings in which the retorts are installed and all other objects exposed to its ravages, is another source of wastefulness as well as a great annoyance and inconvenience. The waste is increased by the admission of air currents to the interior of the retort, tending to cool and harden the surfaces of the wood exposed. This retards penetration and causes unequal distribution of heat and moisture resulting in lack of uniformity in suppleness, and mobility of the fibre, rendering it incapable of yielding to forms without crimping or fracturing.

Direct or Live Low Pressure Steam applied under the same conditions (expansively unconfined), owing to greater energy expended, is rendered less effective and vastly more wasteful than is the use of exhaust steam.

It requires in volume and resultant heat energy approximately six to one, when used in the most perfectly constructed retorts now in general use, with indifferent results, as compared with results obtained with High Pressure Steam Tight Retorts.

Where stock is steamed by the low pressure methods it is never possible to steam the stock uniformly throughout, as it is necessary to oversteam the outside in order to put the inside in proper bending condition, and also from the fact that stock is not all of the same fibres or grain, then when the stock comes to the bender, it either breaks on the outside or buckles on the inside caused by the stock being oversteamed or cooked on the outside and not properly softened on the inside, then it does not dry evenly and has a tendency to not hold the desired bend or shape after being taken out of the forms, also, in many factories there occurs quite a heavy breakage after the stock is taken out of the forms and put in the dry-kiln which is caused by the unevenness in steaming by the low pressure methods.

High Pressure Steam enables manufacturers to use the required pressure for the different kinds of stock and different ages, and this pressure ranges from thirty to forty-five pounds which is easily controlled and adjusted by a special reducing valve. With the High Pressure the pressure drives the heat and moisture to the center of the stock, softening the heart exactly the same as the outside and in fact, giving all the stock throughout, the same steaming regardless of the fibres or grain; then when the stock comes to the bender, the benders can be operated at least one-third faster from the fact that the stock does not require as much end pressure, and the stock conforms to the shape to be bent so much easier, also stock will dry in one-third less time in the straps, and in one-half less time in the dry-kiln, from the fact, the High Pressure does not water soak the stock as does the low pressure; also, there is never any after breakage with high pressure steaming from the fact that the stock has all gotten the same amount of steam.

With the use of high pressure steam you can bend material absolutely green on account that the high pressure breaks up the sap cells, forcing out practically all the sap and acids which will leave your stock in a good dry condition after being bent; this feature alone is worth the price of high pressure Retorts as it enables manufacturers to use their stock in the condition they receive it, without having to pile their lumber or waiting for months for it to dry.

No Escaping Steam or Dripping Water

These Retorts are manufactured with heavy cast iron walls and our special patented quick closing door, and are absolutely steam and water tight, enabling you to place them anywhere in your plant wherever most convenient, and it is not necessary to have stacks or chutes of any kind to take the steam out of your buildings, neither is it necessary to have concrete or special floors of any kind; also, you can place these Retorts close to your benders, enabling your men to turn out much more work per day.

Actual Time Savings and a Greater Production

With the Perfections your benders can go to work in the mornings as soon as they care to, as your watchman can turn the steam on a few minutes before starting time for your benders and your stock is always steamed properly; no guesswork with the high pressure Perfections, also, your men will not be idle any part of the day waiting for stock to steam as by the use of the Perfections they give you a continuous operation.

Your men do not have to work in a blinding steam as after the stock has been steamed and the door opened on the Retort, the stock is free of steam, and it is not even necessary to use hooks of any kind to get the material out of the Retorts as same can be taken out with the bare hands; also by being able to place the Retorts close to your benders, they can handle the material faster from the Retorts to the benders and you would also find that you could run your benders at least one-third faster on account of the material having been steamed properly. You would find that these Retorts would increase your production in your bending department at least 20 per cent.

Fifty Per Cent Less Breakage

These Retorts steam your stock absolutely the same throughout, as can be proven by taking a piece of stock steamed by the Perfections, cutting it into and examining the center of the stock. These Perfections do not oversteam any part of the stock as it all gets exactly the same amount. We have found that the thousands of users of these Retorts have all reduced their breakage losses 50 per cent over their former methods and many have reduced their breakage 90 per cent.

Perfections Thoroughly Tested

All Perfection Retorts are tested to 100 pounds water pressure, then to 100 pounds steam pressure, so that we know positively that every box is perfect in every particular and there is no chance of their giving way after being put in operation. Of the thousands of these Retorts we have in operation in plants all over the country we have never had one show any signs of weakness in any respect. These Retorts can be insured the same as boilers as many firms carry insurance on these Retorts.

Handles All Kind of Work

Perfections are manufactured in different cross sectional sizes, of any desired length, and open at one or both ends, so that they can be furnished for any and all kinds of work. They are being used in every known line of bent work all over the United States and in practically every foreign country and are giving the very best of satisfaction. If you feel you have a hard bending proposition, advise us what it is and we will show you wherein Perfections will solve your problems.

Small Intake Pipe—Economical

The amount of steam consumed by the Perfection is not noticeable on your boilers. These Retorts are absolutely steam and water tight and it only takes enough steam to fill them full and to hold the steam to the required pressure; as the discharge lines are closed before the steam is turned into the Retort. We only use a three-quarter steam line on these Retorts.

As a proof that the Perfections require but little steam is the fact that on stock the temperature of the room, sixty-eight to seventy degrees, there will be only one-half to one gallon condensation from each steaming. This small steam consumption is from the fact that the walls on these Retorts are quite heavy so that the effect of the atmosphere on the heat of the inside of the box is practically nothing. Also, the walls being heavy they hold sufficient heat to keep the stock in perfect bending condition for several hours without resteaming.

Perfections Are Fireproof

These Retorts are absolutely fireproof. We have had quite a lot of them go through fires in which all other machinery was completely destroyed, and we have never lost a single Perfection through any fires. This is in itself, of course, quite an item.

The Perfection Retorts

WHILE the principle of steaming under High Pressure was embodied in the first models of our Retorts, a great many improvements have been made in the style, shape and convenience of our later models, and our long years of experience with high pressure steaming has taught us a great many things about steaming wood that we did not at first know. Today our Retorts are true to their name "Perfection" and the results obtained with the models as shown in this catalogue are almost perfect.

In this catalogue are shown reproductions and photographs of the different styles of our Retorts. We can furnish them as you will note, in different cross sectional sizes, of any desired length, and open at one or both ends.

Study the photo reproductions of our Retorts carefully. Then kindly note the pen picture on the introductory page, showing the open method of steaming. This drawing was made from a photograph taken in a large bending plant, however, the steam clouds were so dense as to make use of the actual photograph impossible and it was necessary to draw over it; no doubt, you have seen this same condition in bending plants and if you are using the same method, let us suggest that you put in one of our Retorts and compare it with your present process. You will note according to the drawing mentioned above, where the workmen is using the exhaust steam or, in fact, any low pressure steam, that he is compelled to stand close by the steam box door to get out the stock. The steam clouds are so dense that he cannot see what he is doing and it is simply a matter of "grab and take" whatever he can get hold of.

Perfections Eliminate Staining

The Perfection method of high pressure steaming eliminates all staining and the stock comes from the Retort clean and bright, in strong contrast to the stock which comes from the old style low pressure leaky steam boxes. The reason we do not have stained stock is from the fact that the walls of our Retorts are made of a composition casting, and by our high pressure method, we eliminate all the acids from the stock which causes the staining. At forty pounds pressure, we have two hundred and eighty degrees of heat and this amount of heat breaks up the sap cells and forces out the acids.

Perfection Standard Retort

Open at One or Both Ends.

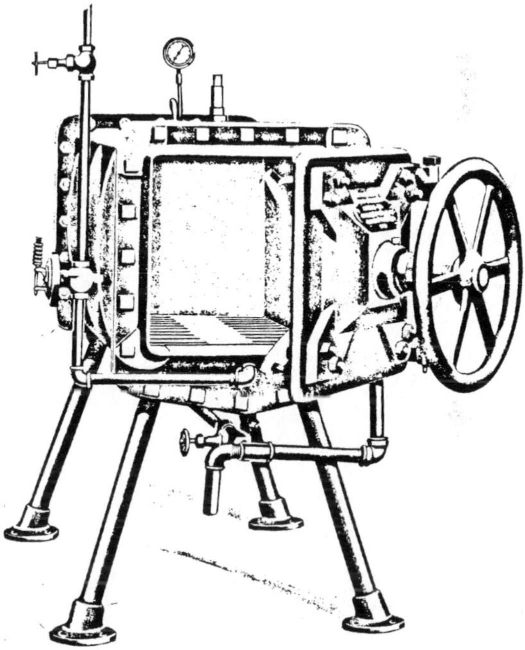

16 Inches Wide, 16 Inches Deep, Any Desired Length.

Perfection Standard Retort

DESCRIPTION

THIS Retort is 16 inches wide by 16 inches deep, door 16 inches square. We can furnish it in stock lengths of 30, 36, 48, 54, 60, 66, 72, 78, 84, 96, 102, 108, 114, 120, 126, 132, 144, 150, 156, 162, 168, 174, 180, 192, 198, 204, 216, 228, 240 and 252 inches, or can make any desired length.

The walls are ⅞ inch thick. The seat forming the closure of the door to the Retort is made of babbitt metal and lead and will last from two to five years.

WE LIKEWISE MANUFACTURE THIS RETORT OPEN BOTH ENDS

Perfection Standard Retort

Open at One or Both Ends.

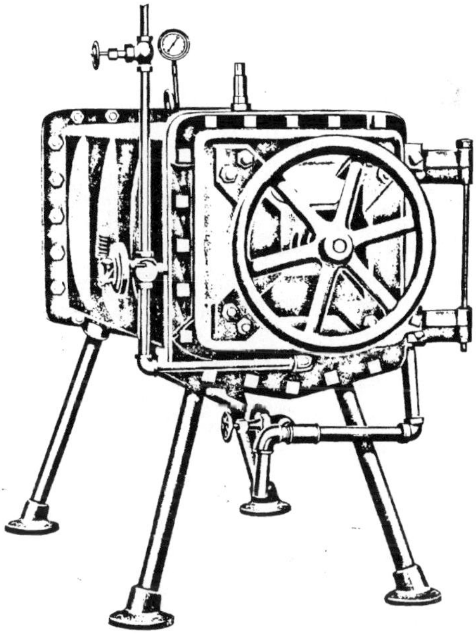

16 Inches Wide, 16 Inches Deep, Any Desired Length.

Perfection Standard Retort

DESCRIPTION

THIS Retort is 16 inches wide by 16 inches deep, door 16 inches square. We can furnish it in stock lengths of 30, 36, 48, 54, 60, 66, 72, 78, 84, 96, 102, 108, 114, 120, 126, 132, 144, 150, 156, 162, 168, 174, 180, 192, 198, 204, 216, 228, 240 and 252 inches, or can make any desired length.

The walls are $\frac{7}{8}$ inch thick. The seat forming the closure of the door to the Retort is made of babbitt metal and lead and will last from two to five years.

WE LIKEWISE MANUFACTURE THIS RETORT OPEN BOTH ENDS

Perfection Jumbo Retort

Open at One or Both Ends.

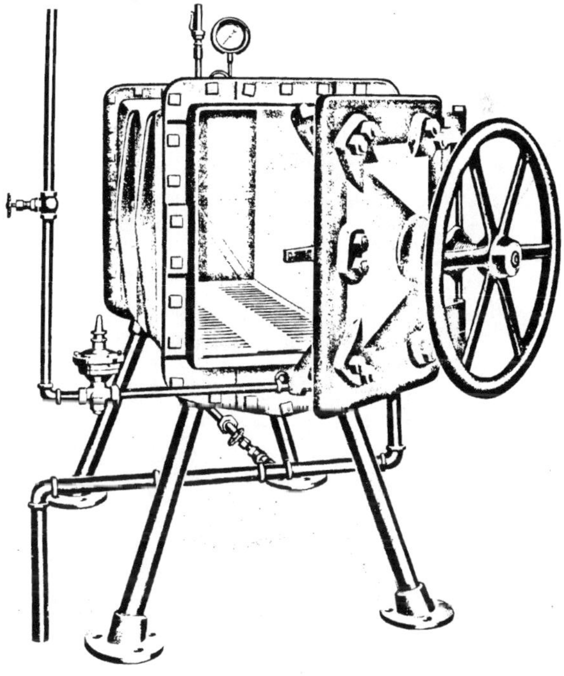

18 Inches Wide, 25½ Inches Deep, Any Desired Length.

Perfection Jumbo Retort

DESCRIPTION

YOU will note by the photograph that this box is reinforced on the sides and door, and made especially strong in every part.

The capacity is large as you can see by interior shown. Please also note the ease with which the door can be opened or closed and locked. With the head of this box we are able to open the door wide and close it again in ten seconds' time.

This is an altogether satisfactory box for heavy work.

Perfection Jumbo Retort

Open at One or Both Ends.

18 Inches Wide, $25\frac{1}{2}$ Inches Deep, Any Desired Length.

Perfection Jumbo Retort

DESCRIPTION

THIS Retort is 18 inches wide by 25½ inches deep, door 18 inches wide by 25½ inches deep. We can furnish it in lengths of 54, 78, 104, 128, 156 inches, or can make any desired length — open at one or both ends. The walls are 1⅛ inches thick.

It has a wonderful capacity. Stock can be piled in this box solid, using no cross strips whatever, and it is large enough for any size stock.

There are firms having these boxes in use for bending stock 4 inches in thickness and 14 inches wide. This box is for use where a large bending capacity is desired or where you use large heavy stock.

Perfection Jumbo Retort

Open at One or Both Ends.

22 Inches Wide, 25½ Inches Deep, Any Desired Length.

Perfection Jumbo Retort

DESCRIPTION

YOU will note by the photograph that this box is reinforced on the sides and door, and made especially strong in every part.

The capacity is large as you can see by interior shown. Please also note the ease with which the door can be opened or closed and locked. With the head of this box we are able to open the door wide and close it again in ten seconds' time.

This is an altogether satisfactory box for heavy work.

Perfection Jumbo Retort

Open at One or Both Ends.

22 Inches Wide, 25½ Inches Deep. Any Desired Length.

Perfection Jumbo Retort

DESCRIPTION

THIS Retort is 22 inches wide by 25½ inches deep, door 18 inches wide by 25½ inches deep. We can furnish it in lengths of 30, 42, 48, 54, 60, 66, 72, 78, 84, 90, 96, 102, 108, 114, 120, 126, 132, 138, 144, 150, 156, 160 inches, or can make any desired length — open at one or both ends. The walls are 1⅛ inches thick.

It has a wonderful capacity. Stock can be piled in this box solid, using no cross strips whatever, and it is large enough for any size stock.

There are firms having these boxes in use for bending stock 4 inches in thickness and 14 inches wide. This box is for use where a large bending capacity is desired or where you use large heavy stock.

The Perfection Guarantee

WE GUARANTEE

To Save You One-half Your Present Breakage, whatever it may be.

WE GUARANTEE

Stock Will Dry in Your Press or Forms in One-third to One-half Less Time than by the old method.

WE GUARANTEE

Stock Will Dry in One-half Less Time in your dry kiln.

WE GUARANTEE

To do Away with Staining of stock in steaming.

WE GUARANTEE

These Retorts to be Absolutely Steam and Watertight. They can be set any place in your factory convenient to your press or forms, thus making a big saving in handling your stock from the Retort to the bender.

WE GUARANTEE

The Steam Consumption to be Small, only taking just enough to fill the box full and to keep the pressure to the desired point.

WE GUARANTEE

To Steam Stock in a Few Minutes' time that would take several hours by any other method, and to steam it uniformly throughout.

WE GUARANTEE

our cast iron Retorts to be Practically Indestructible, and they will never have to be replaced.

WE GUARANTEE

these Retorts will Give Entire Satisfaction.

United States shipments are made from our main factory in Parkersburg, West Virginia.

Canadian shipments are made from our agency at Preston, Ont., saving the Canadian duty on imports.

Some of the Users of Perfections

The Auto Body Co., Lansing, Mich	2 Retorts
The Auto Wheel Co., Lansing, Mich	13 Retorts
O. J. Beaudette Co., Pontiac, Mich	4 Retorts
The Belmar Mfg. Co., Canton, Penn	2 Retorts
Bimel Spoke & Auto Wheel Co., Portland, Ind	10 Retorts
Canadian Shovel & Tool Co., Ltd., Hamilton, Ont. Canada	3 Retorts
The Dayton-Wright Airplane Co., Dayton, Ohio	4 Retorts
The Empire Plow Co., Cleveland, Ohio	2 Retorts
Fisher Body Corp., Detroit, Mich	6 Retorts
Garton Toy Co., Sheboygan, Wis	5 Retorts
Thomas Graham & Co., Madison, Ind	5 Retorts
W. H. Gunlocke Chair Co., Wayland, N. Y	3 Retorts
Hayes Wheel Co., Jackson, Mich	22 Retorts
The Hayes Wheel Co., of Canada, Ltd., Chatham, Ont. Canada	4 Retorts
J. M. Hays Wood Products Co., Jefferson City, Mo	5 Retorts
The Howe Scale Co., Rutland, Vt	3 Retorts
K. & C. Mfg. Co., Henniker, N. H	3 Retorts
Kelsey Wheel Company, Ltd., Windsor, Ont. Canada	4 Retorts
The Knechtel Furn. Co., Hanover, Ont. Canada	2 Retorts
Kurtz Bros., Bethlehem, Pa	3 Retorts
Lansing-Company, Lansing, Mich	3 Retorts
Philip Lebzelter & Son Co., Lancaster, Pa	3 Retorts
Michigan Chair Company, Grand Rapids, Mich	2 Retorts
Milne Chair Company, Chattanooga, Tenn	8 Retorts
Milwaukee Chair Co., Milwaukee, Wis	4 Retorts
Muncie Wheel Co., Muncie, Ind	4 Retorts
The Ontario Wheel Co., Ltd., Gananoque, Ont. Canada	2 Retorts
Parkersburg Chair Co., Parkersburg, W. Va	5 Retorts
Pouvailsmith Corp., Poughkeepsie, N. Y	4 Retorts
Prudden Wheel Company, Lansing, Mich	16 Retorts
Sheller Wood Rim Mfg. Company, Portland, Ind	6 Retorts
The Shreve Chair Co., Union City, Pa	5 Retorts
Sidway Mercantile Co., Elkhart, Ind	2 Retorts
Southern Bending Company, Johnson City, Tenn	6 Retorts
Sturgis Steel Go-Cart Co., Sturgis, Mich	2 Retorts
The Taylor Chair Co., Bedford, Ohio	3 Retorts
The Toledo Metal Wheel Company, Toledo, Ohio	2 Retorts
F. A. Whitney Carriage Co., Leominster, Mass	2 Retorts
Ypsilanti Reed Furn. Co., Ionia, Mich	25 Retorts

ALL CONTRACTS AND AGREEMENTS ARE CONTINGENT UPON STRIKES, FIRES, ACCIDENTS AND OTHER OCCURRENCES BEYOND OUR CONTROL. PRICES SUBJECT TO CHANGE WITHOUT NOTICE.

JERSEY CITY GO-CART CO.
INCORPORATED
MANUFACTURERS OF
BABY CARRIAGES
FACTORY 354-358 MERCER STREET

JERSEY CITY, N.J. June 9th. 1919.

The Perfection Wood Steaming Retort Co.,

Parkersburg, W. Va.

Gentlemen:-

We are in receipt of your favor of June 5th. wherein you request a recommendation as to the saving and improvements we have found by the use of your Retort.

After considerable experimenting with Steaming Retorts of various manufacturers we were fortunate enough to try out your Steaming Retort and after an exhaustive test, wherein we kept an accurate account of splits and breakages, and upon comparing these records with various other machines, we have found that your machine gave us the most satisfactory results, in fact, so much better that there was no comparison; we might also add that the gas consumption of your retort is very light.

We would like very much to have you send us your catalogue as soon as same is completed.

Very truly yours,

JERSEY CITY GO-CART CO. INC.

WM/DT.

No comparison.

H. E. SHELLER PRES. & GEN. MGR. PHIL STACHLER V-PRES. W. C. HIRONS SEC'Y & TREAS.

SHELLER WOOD RIM MFG. CO.
INCORPORATED
MANUFACTURERS OF
AUTOMOBILE STEERING WHEEL WOOD RIMS
BALLBATS BENDING AND TURNING

PORTLAND, INDIANA.

June 6, 1919.

Perfection Wood Steaming Retort Co.,
Parkersburg, W. Va.
Attention Mr. A. R. Adair.

Gentlemen:

We have five of your Retorts in operation in our plant which are giving entire satisfaction and fill all the requirements of your recommendations.

Since installing these Retorts our production has materially increased and the results generally are very satisfactory.

The fact that we just received another Retort from you is very evident that we are pleased with your equipment since we have been using your Retorts exclusively during the last four years.

Yours very truly,

SHELLER WOOD RIM MFG. CO.,

Pres. & Gen. Mgr.

HES/M.

Production increased.

O. J. Beaudette Co.
WHOLESALE MANUFACTURERS OF
AUTOMOBILE BODIES

O. J. BEAUDETTE, PRESIDENT
M. J. BEAUDETTE, V. PRESIDENT
W. W. SMITH, SEC'Y & TREAS.

PONTIAC, MICH. June 6, 1919.

Perfection Wood Steaming Retort Co.,
Parkersburg,
W. Va.

Gentlemen:-

We have your letter of the 4th. in regard to the retorts we have here.

Would say that we have four of your retorts and we have been using them for a considerable time and find them very satisfactory. We believe that the breakage is at least 50% less than the way we were doing it before, and would say that we are entirely satisfied with them.

Yours truly,
O. J. Beaudette Co.

Breakage reduced 50 %.

ESTABLISHED 1864 — INCORPORATED 1901

THOMAS GRAHAM & COMPANY
INCORPORATED
Manufacturers of
VEHICLE WOOD STOCK

J. B. ROSS, PRES'T.
J. G. ROSS, SEC'Y & MGR.

Madison, Ind. June 7th. 1919.

Perfection Wood Steaming Retort Co.,

Parkersburg, W. Va.

Gentlemen:-

Answering your favor of the 5th. inst. will say that we have been using your retorts exclusively for our steaming requirements in bending for the past four years and find them giving excellent satisfaction. We bend all kinds of vehicle rims, both light and heavy, green and dry, and with the proper schedule of steaming for the different sizes and conditions we have little trouble.

We now have in operation five of your machines and as stated above we think they are the best method of taking care of our steaming requirements.

Yours very truly,
Thomas Graham & Co. Inc.
J G Ross Sec'y.

JGR/CW

Bending all kinds of material satisfactory.

FISHER BODY CORPORATION

MANUFACTURERS OF

AUTOMOBILE BODIES

AUTOMOBILE BODY HARDWARE

DETROIT, MICH. June 9, 1919.

Perfection Wood Steaming Retort Co.,

Parkersburg, W. Va.

Mr. A. R. Adair.

Gentlemen:-

We have six of your steaming retorts in use, three of which we have been using for the past five years, and have had excellent results from them.

In these days of improved processes and equipment, it is hardly necessary to make comparison between your retort and the old steam boxes, as the results are so much better and quicker, and the work steamed so much more thoroughly, due to the pressure to which it is exposed. which has always been one of our chief arguments, and which we find works out as you say, that there cannot help but be considerable saving in time, material and steam, together with a more clean and sanitary bending room.

Yours very truly,

FISHER BODY CORPORATION.

R. R. BROWN

REB/MS

High pressure does the trick.

THE HAYES WHEEL Co

OF CANADA LTD

AUTOMOBILE WHEELS HUBS AND RIMS

CHATHAM

June 9, 1919.

The Perfection Wood Steaming Retort Co.,

Parkersburg, W. Va.

Gentlemen:-

Answering yours of June 5th, concerning the experience we had with your retorts for the bending of auto rims. We are pleased to report that they are far superior to the old steam boxes in vogue several years ago, or any other method of low pressure steaming.

Operators on your steam retorts can produce material twice as fast and with less cullage than we have ever experienced before, and if we are contemplating any additional equipment of this nature, you may be sure we will forward our orders to yourselves, as we are entirely satisfied with same.

Yours very truly,

THE HAYES WHEEL CO. OF CANADA, LIMITED.

Wm Kistler

Manager.

Operators can produce material twice as fast.

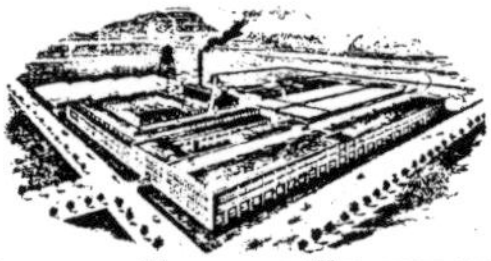

MILNE CHAIR COMPANY

CHATTANOOGA, TENNESSEE
U.S.A.
6/7/19.

Perfection Wood Steaming Retort Co.,

Parkersburg, W. Va.

Dear Sirs:-

We have yours of the 4th, and in reply beg to state we have eight steaming retorts, and we consider them a decided improvement over the old process of steaming; as while we have had very good steam boxes before we put the retorts in, we find there is a saving of at least 50% in breakage over our former method.

If we are needing any additional steaming capacity we would not think of going back to the old steam boxes, after having once used the retorts.

Yours very truly,

MILNE CHAIR CO.

By ----W. S. Milne.

WSM:AGK

Had good Retorts but found 50% saving in breakage by use of Perfections.

MILWAUKEE CHAIR C^o

Makers of Fine Chairs

MILWAUKEE CHICAGO NEW YORK SEATTLE

MILWAUKEE
June 13
1919

Perfection Wood Steaming Retort Co.

Parkersburg, W. Va.

Gentlemen:-

Replying to your letter of the 4th we are pleased to state and it may be used as a recommendation that the Retortswhich has been in use in our factory for some years has given perfect satisfaction. We would not be willing to state just how much this saves over the old style process as we have not even attempted to measure same and have only this to say, that we are thoroughly satisfied with same and have been since it was installed.

Yours very truly,

MILWAUKEE CHAIR COMPANY,

H. Eiff
General Sales Manager

CHEiff
FBH

Thoroughly satisfied.

RANDOLPH CHAIR COMPANY
Chairs and Rockers
DOUBLE CANE AND SPLIT BOTTOM CHAIRS

C. C. CRAWFORD, TREASURER

THE LARGEST DOUBLE CANE FACTORY IN THE SOUTH

ASHEBORO, N.C. 6/7/1919.

Perfection Wood Steaming Retort Co.,

Parkersburg,

W. Va.

Gentlemen:-

We have been using one of your Steaming Retorts for about four years, and we find it almost perfect. In fact we have a very low percentage of breakage.

If we were in the market for another retort, we would not consider making a change in the make.

Yours very truly,

RANDOLPH CHAIR CO.

C. C. Crawford--Treas.

Retorts are perfect.

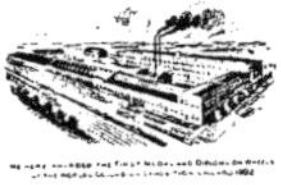

MUNCIE WHEEL CO.

MANUFACTURERS OF

BANNISTER VEHICLE AND SARVEN WHEELS

MUNCIE, IND.
June 7, 1919.

Perfection Wood Steaming Retort Co.,

Parkersburg.

W. Va.

Gentlemen:--

In reply to your letter of June 4th., concerning your steam retorts, would state that they are entirely satisfactory to us and we believe they will show a saving in the use of steam over any other system that we know of.

The best recommendation that we can give is that we are in the market for two more.

Yours very truly,

MUNCIE WHEEL COMPANY

By O. B. Bannister
Prest.

OBB-GEW.

Have ordered two more.

THE ONTARIO WHEEL COMPANY, LIMITED

MANUFACTURERS OF SARVEN & SHELL BAND CARRIAGE WHEELS

GANANOQUE, ONT June 7th, 1919.

Perfection Wood Steaming Retort Company,

Parkersburg, W. Va., U.S. A.

Gentlemen:-

This is to acknowledge yours of the 4th. inst. in which you ask us to advise you what results we are receiving from your Steaming Retorts.

We are pleased to report that we have two of your large size Retorts in use and we are having the best of satisfaction. We would not return to the old system of steaming under any consideration. We save 25% in fuel consumption, 10% in breakage and a greater output with truer rims when they come out of the drying kilns. Also a very important advantage is we are able to bend absolutely green stock with perfect results, which we were not able to do with the old method of steaming.

We are,

Yours truly,

THE ONTARIO WHEEL CO. LTD.

O. Hergne

President & Manager

OH-D

Can bend absolutely green stock.

Kurtz Brothers

Fine Interior Cabinet Work and Special Furniture

Union Street, First and Second Avenues

New York Office
437 5th Avenue

Bethlehem, Pa., U. S. A.

July 9, 1919.

Perfection Wood Steaming Retort Co.,

Parkersburg, W. Va.

Gentlemen:

The Retorts that you have furnished us have given entire satisfaction. We have steamed White Oak for bending purposes which were 4 1/2" x 7 1/2" very satisfactorily and in a very reasonable length of time.

The process that you have of steaming under pressure, we consider far superior over any other methods because we believe that the breakage on wood steamed this way for bending purposes is far less than by any other methods.

Very truly yours,

KURTZ BROTHERS

By Chas. F. Kurtz

1/CES

Note the size material they are bending.

Michigan Chair Co.

Grand Rapids Mich.

June 10th, 1919.

Perfection Wood Steaming Retort Co.,
Parkersburg, W. Va.

Gentlemen;-

The retort you sold us on thirty days' trial in 1908 is all the name implies- Perfection.

It proved such a success that we ordered the large one for long stock. They are both in continuous use. They have cost us nothing for up-keep, and the saving in steam, stock, time and worry has more than paid for the machines.

Perfection is the right name.

Yours very truly,

MICHIGAN CHAIR COMPANY

Jos Garratt

Supt.

FFG F3.

Nothing for repairs since installed in 1908.

LANSING COMPANY

IMPROVED WAREHOUSE TRUCKS.
ELECTRIC INDUSTRIAL TRUCKS.

Cable Address "SUCLANSING, NEW YORK"

Lansing, Mich.

June 6, 1919.

Perfection Wood Steaming Retort Co.,

Parkersburg,

W. Va.

Gentlemen:-

We have had three of your Retorts in operation for the past six years.

They have given us excellent service.

Yours truly,

LANSING COMPANY

M A Templeton

MAT:HGL

Excellent service.

W. O. NELSON

R. M. WATKINS

O. D. HUTCHENS

SOUTHERN BENDING COMPANY

VEHICLE RIMS, HARDWOOD LUMBER
DIMENSION STOCK, SOLID TABLE RIMS
HICKORY LOOM SUPPLIES FOR TEXTILE MILLS
TELEPHONE POLES

JOHNSON CITY, TENNESSEE June 9th 1919

Perfection Wood Steaming Retort Co.,
Parkersburg, W. Va.

Gentlemen:-

Replying to your letter of June 4th.

We are using four 104" retorts and two 128" retorts of your manufacture, and have found them entirely satisfactory in every respect. We find we can steam our stock more uniformly, and in shorter time than we could under our old method and have had no trouble in steaming stock from the lightest buggy rims to the heaviest artillery rims.

Yours very truly,
SOUTHERN BENDING COMPANY
O. D. Hutchens
Secretary

ODH/^

Stock steamed uniformly.

SIDWAY-MERCANTILE CO.
MANUFACTURERS OF BABY CARRIAGES

ELKHART, IND.
June 9th, 1919.

Perfection Wood Steaming Retort Co.,
Parkersburg, W. Va.

Gentlemen;-

Replying to yours of June 5th, regarding recommendation for your retort.

We are not in a position to give you any comparative figures as to the amount of breakage saved by the use of your retort in as much as we have never used any other style except the one manufactured by you and we believe that the first one was purchased about 1913, and the fact that we ordered one since that time is sufficient evidence we are getting complete satisfaction.

We inquired of different wood bending concerns as to the best type of retort before we decided on yours and without exception the concerns we called upon used your make. We are highly pleased with the outfits we have and if occasion arrives we will most certainly take pleasure in recommending your output.

Regretting that we cannot give you a more detailed record of our experience, we are,

Yours very truly,
SIDWAY MERCANTILE COMPANY
Purchasing Agent.

TCW/EW.

Concerns using Retorts say install Perfections.

Sturgis Steel Go-Cart Co.
Manufacturers
Sturgis Michigan

June 6, 1919.

Perfection Wood Steaming Retort Company,
Parkersburg, West Virginia.

Gentlemen:-

It is a pleasure to recommend the use of your Retort for the intricate bendings necessary for our product.

We have gone into this subject very thoroughly on account of the high percentage of breakage and in the face of the rapidly increasing prices of lumber.

Our records show that we are now operating your Retort with LESS THAN 5% BREAKAGE.

We believe this information should be of value to you.

Yours very truly,
C. Wm. Bayne
General Sup't.

CWB/B

Thoroughly satisfied.

YPSILANTI REED FURNITURE COMPANY
Manufacturers of
REED AND REEDFIBRE FURNITURE
Ionia, -:- Mich.
June 24, 1919.

Perfection Wood Steaming Retort Co.,
Parkersburg, W. Va.

Gentlemen:-

Regarding our experience with your Perfection Wood Steaming Retorts, would say that they have been entirely satisfactory in every way. We have in use twenty-two at this time.

As soon as our new bending room, which we have under construction, is completed we will undoubtedly be able to specify for some more.

Assuring you of our appreciation of the manner in which you have always taken care of our orders for new equipment and for the service which your product has given us, we are,

Yours very truly,
YPSILANTI REED FURNITURE COMPANY.

PB/RS.

Have twenty-two and expect to order more.

THE KNECHTEL FURNITURE CO.
LIMITED

HANOVER, ONTARIO

June 9, 1919.

Perfection Wood Steaming Retort Co.,

Parkersburg, W. Va.

Gentlemen:-

Learning that you are getting out a new catalogue, we would ask you to add our Testimonial in regard to High Pressure Steaming Retorts. We have been using your Retorts for seven years, and must say they are giving us entire satisfaction. We would recommend to any one using the old exhaust process to change over to the High Pressure System.

Yours truly,

The Knechtel Furniture Co.,

Per A. Sieling.

LB.

Say to install High Pressure Retorts.

Taylor's Comfortable Rockers

The Taylor Chair Company
BEDFORD, OHIO.

June 14, 1919.

Perfection Wood Steaming Retort Co.,
Parkersburg, W. Va.

Gentlemen:-

In reply to yours of the 5th. beg to advise that we have found your Retort to be satisfactory in every way.

The foreman of our bending room has been doing this work for about 35 years, and is our oldest employee, and he says he is perfectly satisfied with it. We believe that such a recommendation from a man who has been in this position for so long, is the highest possible recommendation.

Yours very truly,

THE TAYLOR CHAIR COMPANY.

BY

FAF:F

Foreman in bending department 35 years, recommends Perfection.

ESTABLISHED 1858 INCORPORATED 1871

F. A. Whitney Carriage Co.

CHILDREN'S CARRIAGES, GO-CARTS, ENGLISH CARRIAGES, PULLMAN SLEEPERS, ENGLISH DOLL CARRIAGES, AND WHITNEY PULLMAN CHAISE (PATENTED)

CABLE ADDRESS WHITNEY
WESTERN UNION CODE
ADDRESS ALL COMMUNICATIONS TO
F. A. WHITNEY CARRIAGE CO

124 WATER ST.

LEOMINSTER, MASS. June 6, 1919.

Perfection Wood Steaming Retort Co.,

Parkersburg, W. Va.

Gentlemen:-

Replying to your favor of June 4th. beg to state that we installed one of your Retorts about one year ago, and it is giving good satisfaction. In fact we have just placed an order for another Retort.

Trusting this information will be of value to you, we are,

Yours truly,

F. A. WHITNEY CARRIAGE CO.

GHR/O

Ordering more of them.

Kelsey Wheel Company, Limited

MANUFACTURERS OF
AUTOMOBILE WHEELS, HUBS, RIMS & STAMPINGS
FINISHERS OF
AUTOMOBILE BODIES.

Windsor, Ontario, June 9th, 1919.

The Perfection Wood Steaming Retort Co.,
Parkersburg, Ont.

Attention of Mr. Adair.

Gentlemen:-

Replying to yours of June 4th, enquiring for our opinion regarding your retorts, wish to say that we have had nothing but the very best of satisfaction with these and would not go back to the old steam boxes for anything.

Your retorts enable us to get the work out much faster and do not take up nearly as much room as the old fashioned steam boxes.

Any time you wish to refer to us, you may do so.

Yours very truly,

KELSEY WHEEL COMPANY LIMITED.

M. G. Campbell

C/MT.

Greater production and a saving in floor space.

THE BELMAR MANUFACTURING CO.
CLOTHING HANGERS
CANTON, PENNA.

June 6,1919.

Perfection Wood Steaming Retort Co.,
Parkersburg, West Va.

Gentlemen:-

Replying to your 4th inst., favor, the two steaming retorts of your manufacture which we have in service are continuing to give satisfaction.

Ease in operation and economical consumption of steam are points in favor of your retorts.-

We have very little breakage.

Very truly yours,

of
THE BELMAR MFG. CO.

LMM/1

Each in operation.

THE DAYTON WRIGHT AIRPLANE CO.
DAYTON, OHIO.
U.S.A. June 6th. 1919.

Perfection Wood Steaming Retort Company.
Parkersburg, W. Va.

Attention Mr. A. R. Adair.

Gentlemen:-

Replying to your letter of June 4th. wish to advise that we have, at the present time, four Perfection Wood Steam Retorts in our plant which have been in operation from two years for the first two to nine and ten months on the last ones purchased.

We have found these retorts an ideal method of steaming airplane longerons, wing bows and the many other hard wood parts which are bent in airplane construction and they greatly conserved our floor space over any other means of steam cooking that we know of. They also give a degree of uniformity which we had not been able to attain in any other device for steaming.

Some of the bendings which we used in your retort were sixteen feet long and required bending at both ends and had been a difficult problem to handle before using your retorts. Under the retort method, the steaming problem for these bendings was solved.

We used ash, several grades of elm and hickory in these bendings and we had very little trouble with wood discolorations which was a source of constant trouble to us in other types of steamers.

If we, at any time, have occasion to increase our facilities for wood steaming for bending purposes, you can be assured that the addition will be your Perfection Retorts.

Very truly yours,
THE DAYTON WRIGHT AIRPLANE COMPANY.
Mfg. Engineer.

EJT-M

Conserve floor space.

SHOVELS, SPADES, SCOOPS, DRAINING TOOLS

Canadian Shovel & Tool Co. Limited.

BRANDS: BULL DOG. FOX. MOOSE. BLACK CAT. CABLE ADDRESS "SHOVEL"

Hamilton, Canada. June 9,1919.

Messrs. Perfection Wood Steaming Retort Co.
PARKERSBURG, W. VA., U.S.A.

Gentlemen:-

You ask us to report on the Steam Retorts which we obtained from you for the purpose of bending our Handles, and in reply we are pleased to say we are very well satisfied with the results obtained from their use.

1st. The Handles are cleaner from stains which formerly came by Hot water heating.

2nd. The Handles dry much quicker.

3rd. We have 50% to 60% saving in culls made by our former method.

The results are very pleasing to us and in building our new Handle Making Shop, we have made provision for further installation of Steaming Retorts.

Yours very truly,
THE CANADIAN SHOVEL & TOOL CO., Limited.

FS/M

Are saving 50 to 60% in breakage.

THE AUTO BODY CO.

LANSING, MICHIGAN

June 14, 1919.

Perfection Wood Steaming Retort Co.,
Parkersburg, W. Va.

Gentlemen:-

Answering your inquiry of recent date beg to say we have no records as to the saving made by the use of your retort. Wish to say, however, that two of them have been in use in our plant five or six years and have been giving the best of satisfaction in every way possible. Your method is far ahead of any other we have tried.

Yours very truly,
THE AUTO BODY COMPANY.
Purchasing Agent.

JBB/H

Our method ahead.

THE NORTH AMERICAN BENT CHAIR CO LIMITED

MANUFACTURERS OF

CHAIRS

OWEN SOUND, CANADA

June 23rd, 1919.

Perfection Wood Steaming Retort Co.,
Parkersburg,
W. Va., U. S. A.

Gentlemen:-

We have your favor of the 4th. inst. with reference to use of the Retort which we purchased from you.

We have had this machine for some years and are very much pleased with it, and would not be without it now.

Yours truly,

The North American Bent Chair Co., Limited

James Garrie
Secretary

A/C.

Would not be without Retort.

... Office of ...

Chas. R. Field Manufacturing Co., Incorporated

John G. Strecker, President, E. W. Strecker, Treas.

Wooden Parts for Children's and Doll Carriages and Go-Carts

Greenfield, Mass., June 10, 1919.

Perfection Wood Steaming Retort Co.,
Parkersburg,
W. Va.

Gentlemen:-

We have used the retort you furnished us with Oct. 1916 about every day since, and find that it does all you claimed for it.

Would also state that we have had no trouble of any kind with it.

Yours truly,

CHAS. R. FIELD MFG. CO.

Doing all we claim.

OFFICE OF

FORT SMITH CHAIR COMPANY

MANUFACTURERS OF

CHAIRS AND ROCKERS

FORT SMITH, ARK.

June 6, 1919.

Perfection Wood Steaming Retort Co.,
Parkersburg, W. Va.

Gentlemen:-

We installed one of your Perfection Wood Steaming Retorts about two and one half years ago, and we have been very much pleased with its operation. We have had no trouble of any kind with it and have been able to do our bending more satisfactory, and with less breakage than ever before.

Yours very truly,

FORT SMITH CHAIR CO.

WA/2

Pleased with Retort.

Outing Kumfy-kab Co.

Originators of Two-Wheeled Folding Baby Cabs. Fully protected by U. S. Patents.

Exclusive Manufacturers of the Outing Kumfykab

Outing Kumfy-kabs. Collapsible and Reed Models. Light, Compact, Comfortable.

La Porte, Indiana, U.S.A.

June 7, 1919.

Perfection Wood Steaming Retort Co.,
Parkersburg,
W. Va.

Gentlemen:

Replying to your letter of June 4th. we cannot give you any figures on this as we have nothing with which to reasonably compare it. Nevertheless, we can say that it is entirely satisfactory and has given wonderful service.

Yours very truly,

OUTING KUMFY KAB COMPANY.

By B. A. Swift
Superintendent.

BTS-CB.

Giving wonderful service.

John D Raab Chair Co
Grand Rapids Mich

June 18, 1919.

Perfection Wood Steaming Retort Co.,
Parkersburg.
W. Va.

Gentlemen:-

The retort which we bought from you four or five years ago, has been very satisfactory in every respect.

Very truly yours.

JOHN D. RAAB CHAIR CO.

JDR CV

John D. Raab....Pres.

Satisfactory in every respect.

IMPERIAL RATTAN CO. LIMITED.
MANUFACTURERS OF
REED, RATTAN AND UPHOLSTERED FURNITURE
STRATFORD. ONT.

June 7, 1919.

Perfection Wood Steaming Retort Co.,
Parkersburg, W. Va.

Dear Sirs:-

We have been using Perfection Retort for several years.

For a large part of our work a retort of this kind is indespensible, saving considerable in both breakage and time.

Yours truly.

HWS/MP

IMPERIAL RATTAN CO. LTD.
Per H.W. Strudley

Retorts indispensable.

SONS-CUNNINGHAM REED & RATTAN CO., INC.
BUILDERS OF
REED FURNITURE OF QUALITY
SINCE 1886
NEW YORK

SPECIAL DESIGNS TO ORDER

June Seventh
1 9 1 9

Perfection Wood Steaming Retort Co.,
Parkersburg,
W. Va.

Gentlemen:-

We are experiencing a great saving in our wood bending department through the use of your Wood Steaming Retort, and we find that it has reduced our breakage to a minimum.

Very truly yours,

Sons-Cunningham Reed & Rattan Co.Inc.
BMB

Breakage reduced to a minimum.

LOUIS F. NONNAST SONS
TABLE MANUFACTURERS
935-1026 N. HALSTED STREET
TELEPHONE DIVERSEY 2277
CHICAGO

June 17, 1919.

Perfection Wood Steaming Retort Co.,
Parkersburg,
W. Va.

Gentlemen:-

In reply to yours of the 4th. inst., we have been using your Retort for some time and are so well satisfied that we have recommended it to numbers of prospective users of wood steaming retorts.

Very truly yours,

LOUIS F. NONNAST SONS.

HWN/IM

Are recommending Retorts.

A. E. CARTIER SONS CO.
LUDINGTON, MICHIGAN

MANUFACTURERS OF
PORCH AND LAWN FURNITURE
AND
SLEIGH GOODS

OFFICERS

June 10, 1919.

Perfection Wood Steaming Retort Co.,
Parkersburg,
W. Va.

Gentlemen:-

We are pleased to state that the Retort purchased of you in 1917 has proven conclusively to be all that you claim for it.

When used According to directions it eliminates practically all breakages.

We can honestly recommend it to anyone having use for a Retort.

Yours very truly,
A. E. CARTIER SONS CO

EFB:LL

By E. F. Burdeau

They are all we claim.

June 12, 1919.

The Perfection Wood Steaming Retort Co.,
Parkersburg, W. Va.

Gentlemen:-

We are in receipt of your favor of the 4th. inst. and in reply the Steam Retort which we installed several years ago is giving every satisfaction, and we find it a great improvement and saving of time over the old way. In fact owing to the lot of work we have continually going through we could not do without it, and can cheerfully recommend same to any concern requiring the use of a Retort.

We are,

Yours truly,
THE H. KRUG FURNITURE CO., Limited
SECRETARY

Could not do without Retort.

The W. H. Mullins Company

MANUFACTURERS OF
AUTOMOBILE BODIES, FENDERS AND OTHER PARTS
IN PRESSED SHEET ALUMINUM AND STEEL
BODIES COMPLETE, INCLUDING WOOD-
WORK FOR OPEN CARS

BK SALEM, OHIO June 11, 1919.

Perfection Wood Steaming Retort Co.,
Parkersburg,
W. Va.

Gentlemen:-

We installed one of your Standard Steaming Retorts about four years ago and it has been giving excellent results during that time, and we do not hesitate to recommend this equipment for the purpose intended.

Yours very truly,
THE W. H. MULLINS CO.
J. H. Blackburn
P. A.

No hesitancy is recommending.

UNION CITY BODY COMPANY
AUTOMOBILE BODIES
UNION CITY, INDIANA

June 10, 1919.

Perfection Wood Steaming Retort Co.,
Parkersburg, W. Va.

Gentlemen:-

ATTENTION: MR. ADAIR.

In reply to your inquiry of June 5th regarding the Retort which we are using: Wish to say that this has been very satisfactory indeed, and were we to install another Bending Outfit, we would certainly call for your Retort.

Assuring you of our appreciation of your services, we are,

Very truly yours,
UNION CITY BODY COMPANY
VICE-PRES. & ASST. GEN. MGR.

Would call for our Retort.

THE HEBBLE-BARLOW CO., INC.
LUMBER MANUFACTURERS
PIEDMONT, ALA.

JUNE
10th
1919

Perfection Wood Steaming Retort Co.
Parkersburg, W. Va.

Gentlemen:-

Yours of the 5th ulto., relative to the satisfaction we have had with the operation of your retort. Beg to advise that the advantages are unlimited. INASMUCH as you have a very liberal trial offer, we see no reason why there should not be an installation of retorts in every Bending Plant in the country.

Please make us quotation on one (1) 14" x 16½" x 74", open both ends. We are now in position to use another.

Yours Very Truly
The Hebble-Barlow Co.
Per John R. Barlow
V. Pres. & Treas.

HBC/JRB

Every bending plant should install.

J. W. Kilgour & Bro. Limited.
Wholesale
Furniture Manufacturers.
Beauharnois, P.Q. June 11th, 1919.

Perfection Wood Steaming Retort Co.,
Parkersburg,
West Virginia, U. S. A.

Gentlemen:-

In reply to yours of the 5th, wish to state that we are very much pleased with your Steam Retort which has been in operation for almost a year. In fact we are so pleased with it that we contemplate installing a second one in the very near future.

We would have no hesitancy about recommending it as the best Steam Retort which we have seen.

Yours very truly,
J. W. KILGOUR & BRO., LIMITED,
Per R. W. Kilgour
Managing Director.

RWK/GH

Are installing another Perfection.

THE W. F. STEWART COMPANY
AUTOMOBILE BODIES
FLINT, MICH.

July Ninth,
1 9 1 9.

Perfection Wood Steaming Retort Co.,
Parkersburg, W. Va.

- Attention Mr. A. R. Adair -

Gentlemen:-

In reply to your letter of the 5th, would advise that we estimate that the Perfection Wood Steaming Retort saves 90% on steam and requires 75% less time to steam the stock.

In addition to this it is a great deal more convenient than the old style steam boxes which we formerly used.

Yours very truly,
THE W. F. STEWART COMPANY,
S. S. Stewart
President.

SSS:RP.

Save 90% on steam.

Peterborough Canoe Co. Limited
Peterborough, Canada June 7/19

Perfection Wood Steaming Retort Co
Parkersburg, W. Va

Gentlemen:

In regard to the retort which we purchased from you over a year ago would say that we find this very satisfactory for bending parts 1½" square or larger, and since we have started to use it we have been able to cut down our breakage to a very small percentage.

Yours truly,
PETERBOROUGH CANOE CO. Ltd.
Mgr.

WAR/E

Breakage cut down.

Murray & Tregurtha Corp.
Manufacturers of HIGH GRADE
INTERNAL COMBUSTION ENGINES
Atlantic, Mass.

June 6, 1919.

Perfection Wood Steaming Retort Co.,
Parkersburg,
W. Va. Attention of A. R. Adair.

Gentlemen:-

The steel Retort which we bought from your Company when building Sea Planes gave us absolute satisfaction. We used it day and night while rushing this work, and always have had very satisfactory results since using it on our boat work.

Very truly yours,
MURRAY & TREGUARTHA CORPORATION.
By John A Murray

JAM:JB

Using same day and night.

Telephone, West 905

...OFFICE OF...

THE B. KLINKER CO.
Manufacturers of All Kinds of
CHAIRS
OFFICE AND WAREROOMS, 818 AND 820 WADE STREET

Cincinnati, O. June June 6, 1919.

Perfection Wood Steaming Retort Co.,
Parkersburg,
W. Va.

Gentlemen:-

In regard to the Steaming Retort that we bought of you, we must congratulate you on having the best, in our opinion, there is on the market, as we have very little breakage in bending our chair tops since we have been using your Steaming process.

Thanking you for same, we remain,

Yours very truly,
The B Klinker Co

Best on the market.

AMERICAN FACTORY
ALBION MICHIGAN

CANADIAN FACTORY
WINDSOR, ONT.

National Spring and Wire Company
Albion, Michigan

June 10, 1919.

Perfection Wood Steaming Retort Co.,
Parkersburg,
W. Va.

Gentlemen:-

We are writing you our appreciation, and incidentially to say a few words on behalf of your Perfection Wood Steaming Retort.

This has been in operation now for the course of 12 months, during which time we have found it to be exceptionally satisfactory and a great labor saver.

Very truly yours,
NATIONAL SPRING & WIRE CO.
J. C. Kampe

FLC--MW.

Great labor saver.

MOUNT AIRY MANTEL AND TABLE COMPANY
HIGH GRADE DINING ROOM AND HALL FURNITURE
MOUNT AIRY, N.C. June 6, 1919.

Perfection Wood Steaming Retort Co.,
Parkersburg,
W. Va.

Gentlemen:-

It affords us pleasure to say that we have been using your retort for some time and it has proven entirely satisfactory. In fact we hardly know how we could get on without it.

Respectfully,
Mt. Airy Mantel & Table Co.
By Calvin Graves, Sec. & Treas.

CG/B

Could not get along without Retort.

THE BUEHNER CHAIR CO.
MANUFACTURERS AND JOBBERS

OFFICE AND FACTORY

EVANSVILLE, IND.
6/9/1919.

To Users of Steam Retorts:-

We have been using a Perfection Wood Steaming Retort for about one year, and it has given entire satisfaction.

We find it is all the Perfection Wood Steaming Retort Company claim for it, and gladly recommend it.

Very truly yours,

The Buehner Chair Co.

All we claim.

J. C. WIDMAN & CO.
Period Dining Room and Bed Room Suites.
Hall Furniture and Mirrors
AIRPLANE FABRICATED PARTS
PANELS AND FUSELAGE

WIDMAN

WIDMAN
AIRPLANE DIVISION

DETROIT Michigan.
June 7, 1919.

Perfection Wood Steaming Retort Co.,
Parkersburg,
W. Va.

Gentlemen:-

We have your letter of June 4th, with reference to Steam Retort we have in our factory. Will say we have been entirely satisfied both with the Retort and your treatment of our company.

The Retort does all the work you claim for it and that is all that is necessary.

If at any time we can be of service to you, and you would like to have any prospective customers write us, we will be glad to recommend the above named item to the best of our ability.

Yours respectfully,

J. C. Widman & Co.
Frank Widman

GLC/W

Our treatment satisfactory.

J. M. FITZGERALD, PRESIDENT
A. H. FLINT, VICE PRES. & GEN. MGR.
W. N. BENNETT, SECRETARY & TREASURER

ADDRESS ALL COMMUNICATIONS TO THE COMPANY

L-W-F ENGINEERING COMPANY, INC.
MANUFACTURERS OF AIRCRAFT AND ACCESSORIES
COLLEGE POINT, LONG ISLAND
NEW YORK

MEMBER OF AIRCRAFT MANUFACTURERS' ASSOCIATION

June 9th, 1919 ......

Perfection Wood Steaming Retort Co.,
Parkersburg,
W. Va.

Gentlemen:-

We have your letter of June 4th, making inquiries as to the success we are having with your Wood Steam Bending Retort, and in reply beg to state that the same has been in continuous operation for about one year, and has worked to our entire satisfaction.

Yours truly,

A. G. Flachbar
General Superintendent
L-W-F Engineering Company, Inc.

AGF-HML

Working to entire satisfaction.

U. S. CHAIR COMPANY
MANUFACTURERS OF
UPHOLSTERED CHAIRS AND ROCKERS

BOARD OF DIRECTORS

CORRY, PENN'A June 9th.,1919.

Perfection Wood Steaming Retort Co.,
Parkersburg, W. Va.

Gentlemen:-

In compliance with your request for a report on a Steam Retort purchased from you will say it gives us great pleasure to say that it has worked to perfection for about one year.

At the time we purchased it we considered the purchase price exorbitant and had a doubt as to it being a wise investment. This doubt has disappeared since we have had the actual experience, and we believe it is one of the most profitable machines we have in our factory, in-as-much as it reduced the breakage, and requires much less time to dry in the press than the old process of the steaming lumber not under pressure.

Yours truly,

U. S. Chair Co.

Most profitable machine in their plant.

EDWARDS CHAIR COMPANY

ROUND POST DOUBLE CANE BOTTOMED CHAIRS A SPECIALTY
HICKORY ROUNDS GLUED IN THE POSTS

June 7, 1919.

Perfection Wood Steaming Retort Co.,
Parkersburg, W. Va.

Gentlemen:-

We have been using Retort bought of you some eight months ago and beg to say that we are more than pleased with the results in using same.

Since we placed Retort in bending room we find that the breakage in bending is such a small percent that it is not worthy of notice, and the time saved by the use of Retort would justify any one in purchasing.

This is the first we have written you since receiving Retort for the reason that we wanted to make several months test before giving you our approval, or otherwise passing on the Retort. The Retort is a saving of money.

With best wishes we are,

Very truly yours,

EDWARDS CHAIR CO., Inc.,

By J. M. Edwards
Sec'y Treas & Manager

Breakage not worthy of notice.

Indianapolis Chair and Furniture Company
of Aurora
MANUFACTURERS OF
DINERS, FANCY CHAIRS AND ROCKERS

Aurora, Indiana. June 7th, 1919.

Perfection Wood Steaming Retort Co.,
Parkersburg.
W. Va.

Gentlemen:-

We have used your retort for about four years and by its use our bent stock comes through so much nicer that we could not think of going back to the old method of steaming and boiling lumber.

We are unable to give you the amount of breakage saved, but it is considerable, as it is very seldom that we break any stock in bending.

Yours very truly,

INDIANAPOLIS CHAIR & FURNITURE CO.

Sec. & Treas.

WJN/T

Stock comes through better.

THE SINCLAIR-ALLEN MANUFACTURING CO. INC.

MAKERS OF THE
—"Sinclair"—
COMMON SENSE CHAIRS

ESTABLISHED 1874 INCORPORATED 1908

June 17, 1919.

Perfection Wood Steaming Retort Co.,
Parkersburg,
W. Va.

Gentlemen:-

Replying to yours of the 5th. The Retort purchased from you has been giving entire satisfaction for several seasons, and has proved practical both in bending and eliminating breakage.

Very truly yours,

THE SINCLAIR-ALLEN MFG. COMPANY.

A practical Retort.

HAYES WHEEL COMPANY
MANUFACTURERS OF
AUTOMOTIVE WHEELS COMPLETE
WOOD-WIRE-FABRICATED STEEL

JACKSON, MICH. U.S.A.

June 18th, 1919.

Perfection Wood Steaming Retort Co.,
Parkersburg, W. Va.

Attention of Mr. A. R. Adair

Gentlemen:-

Replying to yours of the 4th inst., we are now using at our different plants, twenty-two of your improved retorts. The writer is pleased to say that they are giving very good satisfaction.

With these Retorts, we are able to steam our strips in about twenty-five minutes, instead of giving them from four to five hours in the old way with steam boxes. We are also able to save considerable time in kiln drying the bent rims, over the time required for this purpose when the rims are steamed the old way. We find that these Retorts need little or no attention as there is nothing whatever about them to get out of order, and we would not consider, for a moment, going back to the old way of steaming our strips.

Very truly yours,

HAYES WHEEL COMPANY.

Vice President.

WCM:KM

Nothing to get out of order and need no attention.

KLOPSTOCK BROS.

TELEPHONE MARKET 724

SAN FRANCISCO

June 26, 1919.

Perfection Wood Steaming Retort Co.,
Parkersburg, W.VA.,

Gentlemen:-

Replying to your communication of the 5th inst. relative our opinion of your Wood Steaming Retort, we are glad to say, after using same for one year, that it has done all that you guaranteed it to do. Not only this, but we have, by it's means, increased our output in the bending line, and our only regret is that we did not have it years ago. The cost of the equipment is lost, in our opinion, in the saving that it has made for us, both in time and expense of this work.

We would be pleased to recommend same to anyone who contemplates bending of any nature.

With very best wishes, we beg to remain,

Very respectfully yours,
KLOPSTOCK BROS.

FK:RL

Doing all we Guarantee.

THE FULTON COMPANY

MANUFACTURERS OF

FULTON FOLDING GO-CARTS

CHILDRENS CARRIAGES AND FULTON SERVICE WAGONS

FACTORY AND GENERAL OFFICE BAY CITY MICH

BAY CITY, MICH.

June 6, 1919.

Perfection Wood Steaming Retort Company,

Parkersburg,

W. Va.

Gentlemen:-

For your information we have used one of your Steam Retorts for about two years. We can say that we have never had a bit of trouble with it from the time we started to use it, and it has been in use every day.

We fail to see how anyone who forms wood in any shape can get along without a Steam Retort.

Very truly yours,

THE FULTON COMPANY.

AJA/EC.

Trouble proof.

C. G. CONN LTD.

MANUFACTURER OF HIGHEST GRADE MUSICAL INSTRUMENTS

June 28, 1919.

Perfection Wood Steaming Retort Co.,
Parkersburg, W. Va.

Gentlemen:

The Perfection Steaming Retort which we installed about a year ago has proven absolutely satisfactory. The old method of steaming is not to be compared and the saving on our bendings in many cases advance as high as fifty per cent. It fully warrents all your claims. We could not do without it, we are

Yours very truly,
C. G. CONN LTD.
Per.

EJO/HH

Saving 50% on breakage.

DEFOE BOAT AND MOTOR WORKS

BOATS FOR ALL PURPOSES

BAY CITY, MICHIGAN

March 20, 1919.

Perfection Wood Steaming Retort Co.,

Parkersburg,

W. Va.

Gentlemen:-

Your letter of March 19th, asks us to buy one of your wood steam retorts. We will give you a testimonial instead. We bought one of these retorts about a year ago. It is everything that you claim for it. It saves a big loss in breakage, and a man does, not 20% more work per day, but 50% more per day, and more than that, the work stays where it is put.

Yours truly,

DEFOE BOAT & MOTOR WORKS.

H/H. J. Defoe.

Increases output 50%.

THE KELLER MANUFACTURING COMPANY
INCORPORATED
MANUFACTURERS OF FARM WAGONS
CORYDON, INDIANA.

June 9, 1919.

Perfection Wood Steaming Retort Co.,
Parkersburg,
W. Va.

Gentlemen:-

In reply to your inquiry of the 4th.

We are glad to report that we have had in use for several years one of your Wood Steaming Retorts in connection with our Bending Plant, and we have found same entirely satisfactory in every respect, and a great improvement over the old type steam box for steaming bending strips.

Yours very truly,

The Keller Manufacturing Co.,
By E. G. Keller Mgr.

EGK:B

Retort entirely satisfactory.

VEAL REED CHAIR CO.
Manufacturers of
QUALITY REED FURNITURE
648 Belgrave Ave.
HUNTINGTON PARK, CALIFORNIA

June 11, 1919.

Perfection Wood Steaming Retort Co.,
Parkersburg,
W. Va.

Gentlemen:-

We have used your Retort for over two years and find it very satisfactory in every way.

Yours very truly,

VEAL REED CHAIR CO.
PER J. R. Veal

It is O. K.

The General Woodwork Co.

MANUFACTURERS OF
General Woodwork
AND
Automobile Specialties

PHONE WEST 1778
NUMBERS 1225-1235
BUDD STREET
CINCINNATI, OHIO

July 2nd, 1919.

The Perfection Wood Steaming Retort Co.,
Parkersburg,
W. Va.

Gentlemen:-

It pleases us to inform you that your Retort, now in use for more than a year, has proven satisfactory in every respect.

We have found the same a saving and an improvement on any methods heretofore employed; in fact, we do not hesitate in saying

"Your Retorts Pay For Themselves"

Yours very truly,

PER AFT. THE GENERAL WOODWORK COMPANY

Retorts pay for themselves.

F. H. Conant's Sons, Inc.
Established 1851 Chair Manufacturers Incorporated 1915
Camden, New York.

June 9, 1919.

Perfection Wood Steaming Retort Co.,
Parkersburg,
W. Va.

Gentlemen:-

The retort made by you which we are using in our bending room is proving very satisfactory, and we believe it is an improvement over any other method we have used.

Yours very truly,

F. H. Conant's Sons, Inc.

Improvement over any other method.

THE AUTO WHEEL COMPANY
MANUFACTURERS OF AUTOMOBILE WHEELS

LANSING, MICH.

June 7, 1919.

Perfection Wood Steaming Retort Co.,
Parkersburg,
W. Va.

Gentlemen:-

Replying to your letter of June 4th, inviting criticism or praise regarding your Retorts, have to say the fact that we are now installing three new Retorts in addition to the ten which we have been operating tells the story. If we did not believe them efficient in every way for our work you may rest assured we would be replacing rather than adding to our equipment.

Very truly yours,
THE AUTO WHEEL COMPANY,
By: M. M. Ford
Asst. Secretary

MMF L

Additional installation tells the story.

GIBBS GAS ENGINE COMPANY
OF FLORIDA
SHIP AND ENGINE BUILDERS

JACKSONVILLE, FLORIDA June 9th, 1919.

Perfection Wood Steaming Retort Co.,
Parkersburg, W. Va.

Gentlemen:-

~~Answering your letter of the fourth instant.~~

Answering your letter of the fourth inst:

We used your retort in bending the timbers for the sixteen submarine chasers which we built for the U. S. Navy. We cannot speak too highly of this retort. It gave us absolute satisfaction and we are continuing to use it in all our small boat work.

Will be glad to have you refer anyone to us at any time.

Yours very truly,
GIBBS GAS ENGINE COMPANY.
Vice President.

Cannot speak too highly.

Lenoir Chair Manufacturing Co.

Lenoir, N.C.
June 7, 1919.

Perfection Wood Steaming Retort Co.,
Parkersburg,
W. Va.

Gentlemen:-

Replying to yours of the 4th, would advise that the retort which we purchased from you has given very good satisfaction, and we could not do without the same, and in case we were going to order another of these retorts would first take the matter up with you.

Yours truly,
Sec. & Treas.

J/T

Could not do without Retort.

ALEXANDER McDONALD, INC.
SHIP BUILDER
PORT RICHMOND
STATEN ISLAND

NEW YORK June 16, 1919.

Perfection Wood Steaming Retort Co.,
Parkerburg, W. Va.

Gentlemen:

Replying to your letter of June 4th, asking how we are pleased with the wood steaming retort purchased from you about a year ago.

I want to say that we have kept a careful record of the work that has been handled through this retort and find that it has cut down our breakages 90% and is without exception the best wood steaming retort I have ever seen.

Very truly yours,
ALEXANDER MCDONALD, INC.
Alexander McDonald
President.

Reduced breakage 90%.

GARTON TOY CO.

MAKERS OF GOOD TOYS SINCE 1879

MANUFACTURERS OF

THE BADGER LINE OF BOYS' WAGONS, AUTOMOBILES, VELOCIPEDES, HAND CARS, SLEDS, ETC.

SHEBOYGAN, WISCONSIN June 7, 1919.

Perfection Wood Steaming Retort Co.,
Parkersburg, W. Va.

Gentlemen:--

Replying to yours of June 4th relative to our experience with the use of Perfection Wood Steaming Retorts.

We first put in one only of your retorts, a single end machine, a number of years ago. Since this time we have added to this equipment until we have today in operation five of your retorts, and in fact, our bending department is now equipped with nothing else for steaming stock preparatory to bending except your machines.

We have fully convinced ourselves that this system is the most economical that has come under our observation.

Yours respec GARTON TOY CO.

R. L. Whitehill
SECT. AND TREAS.

RLW:FSK

Additional installation their best recommendation.

K & C MANUFACTURING CO.

MANUFACTURERS OF

K & C

LOCK JOINT WOOD RIMS

Henniker, N. H., June 17, 1919.

Perfection Wood Steaming Retort Co.,
Parkersburg, West Va.

Gentlemen:

We are in receipt of your letter of June 4. In reply will say that we have three of your steam retorts at work in our factory all the time, night and day. They give perfect satisfaction and by using them the result is a saving of 25% over our old process of steaming hard rock maple.

Yours truly,

K & C Manufacturing Co.,
Clinton Lovell. Gen. Mgr.

CL/L

Perfect satisfaction.

Bimel Spoke and Auto Wheel Company

MANUFACTURERS OF HIGH GRADE SPOKES AND AUTOMOBILE AND MOTOR TRUCK WHEELS

Portland, Indiana

June 10, 1919.

Perfection Wood Steaming Retort Co.,
Parkersburg, W. Va.

Gentlemen:-

Answering your inquiry of the 4th. we have in use ten of your retorts. We have found these very satisfactory and get considerably more stock bent than we would by any other method. We find that these get much better results on Hard Maple and Hickory than our previous methods, and find that they are alright for bending Oak, providing the work is not rushed too much, which of course, holds true in bending Oak by any other method.

We get stock thru the dry kilns in quicker time by using the retorts than we did by the old methods. We are glad to recommend this machine.

Yours very truly,

BIMEL SPOKE & AUTO WHEEL CO.

Pres & Gen'l Mgr.

CB.F

Can bend more material and quicker drying.

J. M. HAYS WOOD PRODUCTS CO.

SADDLE TREES, PACK TREES, STIRRUPS

JEFFERSON CITY, MO.

June 9, 1919.

Perfection Wood Steaming Retort Co.,
Parkersburg,
W. VA.

Gentlemen:-

Answering yours of June 5th, cannot give you satisfactory reply, because we have not used these retorts long enough. But will say, from the time they have been put in they have been exceedingly satisfactory, much better than the old ones, and would not do without the outfit for three times the price.

Yours very truly,
J. M. HAYS WOOD PRODUCTS CO.

JMH/AC

They are worth three times their price.

GRANT E. SMITH, President
ELIAS C. VAIL, 1st Vice-President
JOHN VAN BENSCHOTEN, 2d Vice-President and Sales Manager
DR. J. WILSON POUCHER, Treasurer
HENRY S. MORGAN, Secretary and Factory Manager

POUVAILSMITH CORPORATION
POUGHKEEPSIE - NEW YORK

STEERING WHEELS
RADIATOR FILLER CAPS
GEAR SHIFT LEVER BALLS
MOULDED ELECTRIC INSULATION
SPECIAL COMPOSITION MOULDINGS

TRADE POVASCO MARK

June 25, 1919.

Perfection Wood Steaming Retort Co.,
Parkersburg, W. Va.
Attention Mr. W. R. Adair.

Gentlemen:

Replying to your letter of the 16th. in reference to the results we are getting from the Perfection Wood Steaming Retorts, we take great pleasure in advising you that the results obtained by this method of steaming are far in excess of our expectations. While we were negotiating with you on these Retorts we were very positively advised that this method of steaming was worthless for the bending of hard maple and that while they would give satisfactory results on oak and hickory, this was the limit of their capacity. The writer has had many years experience in the bending of hard white maple in the manufacture of bicycle wood rims and automobile steering wheels and has used a variety of steaming apparatus for this purpose, but we say to you frankly that we have never used any method of steaming maple which will in any way compare for excellence of results that we are getting today with your Perfection Wood Steaming Retorts. We carefully test each carload of stock when it reaches us and ascertain the percentage of moisture the lumber contains and we steam it accordingly. Operating the Retorts in this manner we are bending half circles in a Defiance felloe bender, ranging from 16" to 22" outside diameter, and our loss is practically nothing. We do not get one broken back out of a thousand bendings and we are able to dry the bendings within four or five days. In our estimation it would be impossible to accomplish any better results than your Retorts are giving in every day use. There is absolutely no guess work in using them and when we make a thousand bendings we know absolutely that we shall get very nearly one thousand good bendings when they come out of the dry kilns.

It is a great pleasure to us to write you as above and as we enlarge our bending capacity, as we certainly shall, you can rest assured that there will be no Retorts used in our factory other than the Perfection.

With the writer's kind personal regards to your good self, we beg to remain

Very truly yours,

POUVAILSMITH CORPORATION

Karl Hodges

KH/D

Breakage loss on steering wheels 1/10 of 1%.

FITZ GIBBON & CRISP, INC.
AUTOMOBILE BODIES
TRENTON, N.J.

June 6th, 1919.

Perfection Wood Steaming Retort Co.,
Parkersburg, W. Va.

Gentlemen:-

Attention Mr. A. R. Adair.

Replying to yours of the 4th, in reference to the use of the Steaming Retort which we purchased from you sometime ago, beg to advise that your machine has been giving us very satisfactory results, and we find it very practical and a very great improvement over our former method of bending. We could recomend it to anybody in our line who attempts to do their own bending.

Trusting that we may continue in the successful operation of your machine, we beg to remain,

Yours very truly,

FITZ GIBBON & CRISP, Inc.

LLW-CLB.

Would recommend Retort.

ESTABLISHED 1856

PHILIP LEBZELTER & SON CO.

EAGLE WHEEL AND BENDING WORKS

LANCASTER, PA. 6/9/19.

Messrs. Perfection Wood Steaming Retort Co.,
Parkersburg, West. Va.

Gentlemen:-

We are in receipt of yours of the 4th. inst.. and in reply we are pleased to say we have three of your Retorts at the present time, and they are giving us entire satisfaction both as to saving of time and the small percentage of breakage in bending, and we can safely recommend them very highly to anyone, as we for one would use no other Retort excepting your make, and when we replace our old steam boxes we always do so with your equipment.

Yours truly,

PHILIP LEBZELTER & SON COMPANY.

CJL/HCL.

Would not use any other Retort.

THE SHREVE CHAIR COMPANY
INCORPORATED
MANUFACTURERS OF
CHAIRS AND ROCKERS

UNION CITY, PA. June 7, 1919.

Perfection Wood Steaming Retort Co.,
Parkersburg, W. Va.

Gentlemen:-

We still have the five Wood Steaming Retorts that we bought from you several years ago. They are all in use in our bending room, and are doing as good work as ever.

There seems to be no wear out to them.

Yours truly,

THE SHREVE CHAIR CO.

L. D. Shreve

LDS/M

Doing as good work as when installed in 1911.

PRUDDEN WHEEL COMPANY
MANUFACTURERS OF
AUTOMOBILE WHEELS COMPLETE

LANSING, MICH. June 7th, 1919.

Perfection Wood Steaming Retort Co.,
Parkersburg, W. Va.

Gentlemen:-

ATTENTION MR. A. R. ADAIR.

We have installed a total of sixteen of your Retorts for the steaming of rim strips for bent rims. These have been giving excellent and satisfactory service in the bending of oak, hickory and heavy rim strips. We do not hesitate to recommend them highly to any manufacturer of bent rim stock.

Yours very truly,

PRUDDEN WHEEL COMPANY.

Dict. by H. F. Harper.

Satisfactory service.

The Toledo Metal Wheel Company

Children's Vehicles

Toledo, Ohio June 20, 1919.

Perfection Wood Steaming Retort Co.,

Parkersburg, W. Va,

Gentlemen:-

In reply to your letter of June 5th, wish to state that the Wood Steaming Retorts which we have purchased from you have been giving us excellent satisfaction, in fact we are frank to say that they bear out all the claims which are made for them.

We are able, not only to get out a much larger output with these retorts than with anything we have formerly used, but we also find that the wood is much more thoroughly steamed and we are able to make severe bends with a much smaller percentage of breakage.

We do not hesitate to heartily recommend them to anyone who has use for steaming retorts.

Very truly yours,

THE TOLEDO METAL WHEEL CO.

JW L.

They do all we claim.

HOCH HAMES CO. INC.

MANUFACTURERS OF HIGH GRADE HAMES

FREELAND, PA. June 25, 1919.

Perfection Wood Steaming Retort Co.,

Parkersburg,

W. Va.

Gentlemen:-

We are very glad to report that the Retort which we purchased from you a number of years ago has proven very satisfactory.

We cannot recommend it too highly, for by using the retort there has been a great saving in the steaming of hame wood.

Yours very truly,

THE HOCH HAMES COMPANY, Inc.

GBH/IB

G. B. Hoch Pres

Great saving in steam.

THE UNIT CONSTRUCTION COMPANY

58TH STREET AND GRAYS AVENUE

PHILADELPHIA

June 6th, 1919.

Perfection Wood Steaming Retort Co.,
Parkersburg, West Virginia,
Attention Mr. A. R. Adair.

Gentlemen:

Permit us to take this opportunity of going on record with our endorsement and approval of your Perfection wood steaming retorts. In connection with our airplane and flying boat production for both Army and Navy during 1918, we had occasion for constant use of the special 24ft. retorts which you made for us, and secured most excellent results therefrom from standpoint of both economy and efficiency.

Your very prompt delivery of these retorts in the special size required, was also an exceptionally good illustration of the service which you are in a position to give to any concern having requirement for beam bending equipment and to any such concerns we will at all times be very glad to endorse your retorts without qualifications.

Yours very truly,
Unit Construction Company

Rayburn Clark Smith

President.

RCS/H

Our delivery satisfactory.

BLOCH GO-CART CO.

BLOCH

BABY CARRIAGES
INVALID AND ROLLING CHAIRS

1134-1148 AMERICAN ST., NORTH

PHILADELPHIA, PA., U. S. A. 6/16/19

Perfection Wood Steaming Retort Co.,
Parkersburg,
W. Va.

Gentlemen:-

In reply to your inquiry of the 4th, beg to advise that having lately only installed one of your Retorts we cannot in justice comment on its merits at this time. We will, however, say that the Retort has proven more satisfactory than the previous installation we had in use.

The majority of our Bending is being done by outside sources. We are only doing the special work, hence it is rather difficult for us to keep a line on any savings and improvements over our old methods. We feel, however, as the men are progressing with the work that they will find the Retort more advantageous to use than the old system.

In event of our desiring to increase our capacity, or take on and do all our own work, we will gladly consider placing in additional Retorts of your make.

Yours very truly,
BLOCH GO-CART CO.

ARL-HMW.
D.

Retorts a help.

Spotted something that isn't quite right?

Let us know and we'll fix it for future reprints. And as a thank you for your diligence, we'll send you a free Sothis book of your choice.

publisher@sothis.press

This page intentionally left blank

Except of course for that line above. And this one.
And whilst we have your attention, perhaps you'd
like to follow us on Instagram @sothis.press. Go on.